J 523.45 /HAU

Y0-CUL-181

WITHDRAWN

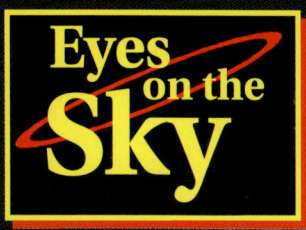

Jupiter

by David M. Haugen

KidHaven Press, an imprint of Gale Group, Inc.
10911 Technology Place, San Diego, CA 92127

Library of Congress Cataloging-in-Publication Data

Haugen, David M.
    Jupiter/by David M. Haugen.
       p. cm. — (Eyes on the Sky)
    Includes bibliographical references and index.
    Summary: Discusses the planet Jupiter including its discovery by Galileo, its location in the solar system, its composition and appearance, its many moons, and NASA's efforts to study it.
    ISBN 0-7377-0716-X (lib. : alk. paper)
    1. Jupiter (Planet)—Juvenile literature. [1. Jupiter (Planet)] I. Title. II. Eyes on the Sky (San Diego, Calif.)
    QB661 .H38 2002
    523.45—dc21

    00-012816

Copyright 2002 by KidHaven Press, an imprint of Gale Group, Inc.
10911 Technology Place, San Diego, CA, 92127

No part of this book may be reproduced or used in any other form or by any other means, electrical, mechanical, or otherwise, including, but not limited to, photocopying, recording, or any information storage and retrieval system, without prior written permission from the publisher.

Printed in the U.S.A.

Table of Contents

Chapter 1
Ruler of the Planets 4

Chapter 2
The Cloud Planet 13

Chapter 3
The Rings of Jupiter. 22

Chapter 4
The Moons of Jupiter. 30

Glossary 40

For Further Exploration 43

Index 45

Picture Credits 48

1
Ruler of the Planets

Jupiter is the largest planet in the **solar system**. The distance across the planet's center—or its diameter—is 88,736 miles. That is eleven times the size of Earth's diameter. If Jupiter was hollow, it could hold more than one thousand Earths. In fact, all the matter that makes up all the other planets in the solar system could easily fit within Jupiter. Because of its impressive size, the planet was named after the chief god of Roman mythology. The god Jupiter (or Jove) was the ruler of the skies to early Romans, and the planet which bears his name seems to lord over the other eight planets in the solar system.

The solar system is made up of the sun and the nine planets that revolve around it.

Jupiter is the fifth planet from the sun, while Earth is the third. Jupiter is about 480 million miles from the sun, or about five times as far away as the Earth is. This also means Jupiter is about 390 million miles from Earth. Still, Jupiter is so large that it can be seen in Earth's sky despite being so distant. In fact, Jupiter is the fourth brightest object in the sky, after the sun, the moon, the planet Venus, and occasionally Mars.

The Romans believed the god Jupiter was the ruler of the skies.

They like

The Giant Gas Giant

Although it is so far away, Jupiter is the closest of the planets known as gas giants. These planets—Jupiter, Saturn, Neptune, and Uranus—do not have hard, rocky surfaces like Earth or Mars. Instead, these gas giants are probably made up almost entirely of gases held together by **gravity**. Some, including Jupiter, may have a rocky core, but astronomers cannot say this for certain since they have nothing that can peer through the layers and layers of thick gases that hide the interiors of these planets.

Since Jupiter is the largest planet in the solar system, it is also the largest gas giant. But Jupiter is about as large as a gas planet can be. Because of its size and mass, Jupiter has a very strong gravity; if any more matter was added to Jupiter the gravity would compact it. This would keep the planet from significantly expanding. Only if Jupiter became a star could it get significantly larger, and that would mean adding eighty times more mass to the planet.

A Planet Like the Sun

Oddly enough, many early astronomers thought Jupiter was a star like the sun. In 1610, Galileo Galilei, a famous Italian astronomer, pointed one of the first telescopes

The Nine Planets

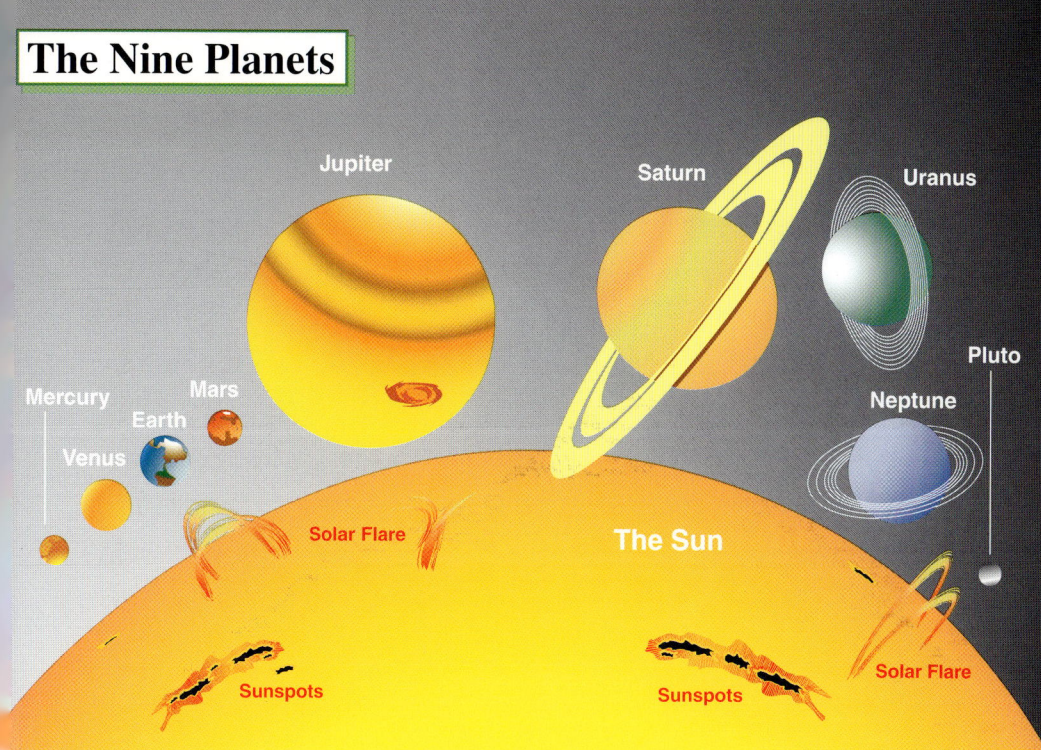

at Jupiter and caught sight of four heavenly bodies circling the planet. Although modern scientists know these were four of Jupiter's sixteen moons, at that time Galileo believed them to be planets **orbiting** what he thought was a small sun. In fact, before the high powered telescopes of the twentieth century, many astronomers carried on a belief that Jupiter was another, distant sun.

Even today, Jupiter's size leads scientists to believe that it almost became a star. The sun became a star because there was enough matter—dust and gases—compressed into a small area that pressures and temperature rose high enough to create explosive nuclear reactions.

Ruler of the Planets   **7**

Galileo pointed his telescope at Jupiter, mistaking it for another sun.

When Jupiter was formed, there was not enough matter to start these types of reactions. Still Jupiter radiates heat, more than twice as much as it receives from the sun.

Jupiter is also like the sun in its composition. Like the sun, Jupiter's atmosphere is made up of two main gases, **hydrogen** (90 percent) and **helium** (10 percent). There are also traces of **ammonia**, **methane**, water vapor, and solid rocky compounds. This is very close to what scientists believe existed in the **solar nebula** that gave birth to the sun and the planets billions of years ago. Thus, astronomers study Jupiter, in part, to learn more about how the entire solar system was formed.

A Huge, Spinning Magnet

Unlike the sun, however, Jupiter rotates quite rapidly. The sun takes roughly a month to spin once on its axis, while on Jupiter a day is only about ten hours long. This is the fastest

Jupiter is the largest gas giant in the solar system.

rotation period of any planet. The quick spinning motion squashes the planet a bit so that the equator actually bulges. This bulge is very noticeable, as the distance from the center of the planet to the equator is roughly 10 percent greater than the distance from the center to the north or south pole.

Because Jupiter is so big and spins very quickly, the planet has an unusually large magnetic field. Each planet has a magnetic field—called a **magnetosphere**—that projects out into space. Jupiter's magnetic field is stronger than that of any other planet. It also extends very far out from the planet, but not

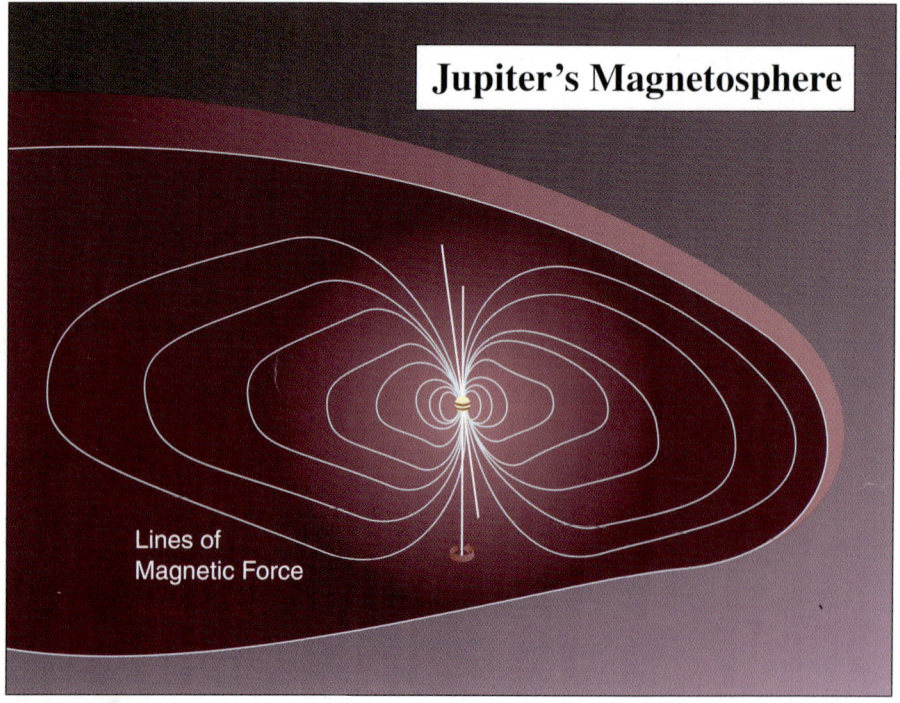

equally on all sides. Jupiter's magnetosphere is elliptical, or oval shaped. It reaches between 2 and 4 million miles toward the sun, but in the direction opposite the sun, it extends over 460 million miles, bringing it within Saturn's orbit.

The magnetosphere attracts charged (thus, magnetic) particles toward Jupiter. Many of these charged particles exist as radiation that surrounds the planet. Earth has similar radiation belts surrounding it, but since Earth's magnetosphere is weaker, fewer particles are trapped in this radiation field. Around Jupiter, however, the powerful magnetic field draws many of these particles. The radiation is so intense surrounding the planet that it would be immediately fatal to unprotected humans. It even interferes with instruments on the spacecraft sent to study the planet.

A Mysterious Planet

Since few Earth probes have reached Jupiter, much of the planet remains a mystery to scientists. One way scientists try to learn more about Jupiter is by using what they know about the planet and conducting experiments on Earth to simulate what might be happening on Jupiter. For instance, knowing that the hydrogen in Jupiter's interior is under tremendous pressure, scientists, working in laboratories, have subjected hydrogen to

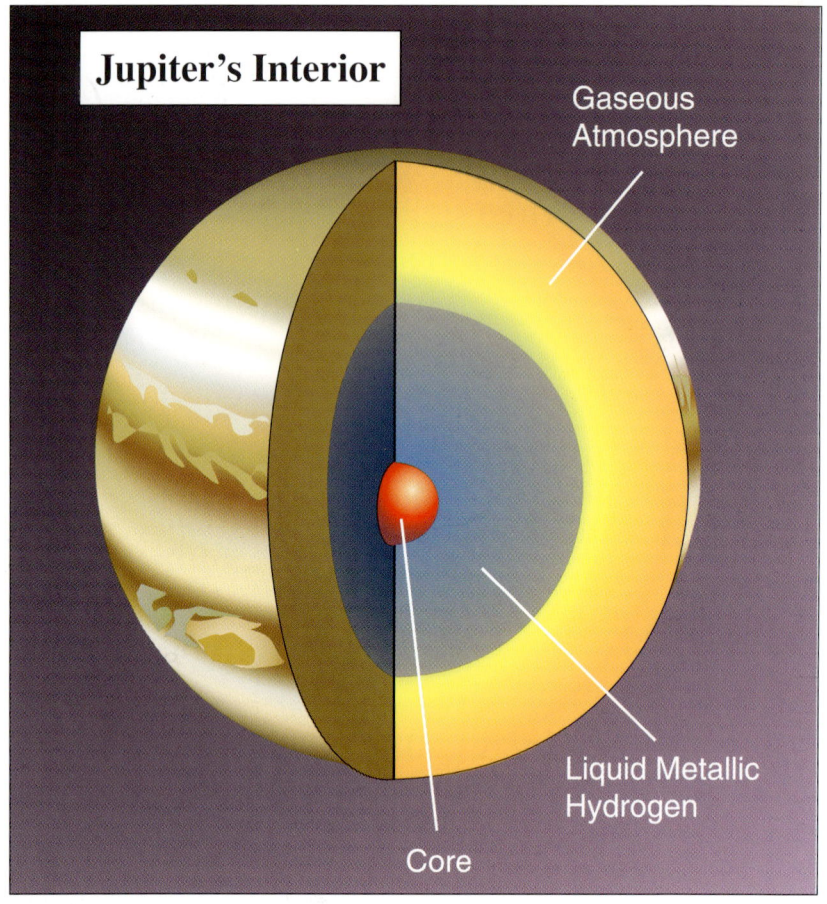

similar pressures and found that it turns from a gas into a liquid metal. Scientists, then, safely assume that there is a large ocean of this liquid hydrogen that rests below the gaseous atmosphere. Such guesswork is how scientists will probably always learn about the interior of Jupiter, because the pressure there is so great that no man-made probe could survive to see the liquid ocean or the rocky core, if they exist at all.

2
The Cloud Planet

Although much of what scientists know about Jupiter is based on tests done in a laboratory, astronomers have always been able to learn many things about the planet by studying its outward appearance. The visible face of Jupiter appears as bands of color with circular regions of varying size wedged in between some of the bands. These bands and circles are cloud formations that fill the planet's upper atmosphere. Scientists believe that the clouds form at least three different layers and may extend downward for thousands of miles.

Cloud Bands

Unlike clouds on Earth which consist of water droplets, the cloud layers on Jupiter are made up of water, ice, ammonium hydrosulfide, and ammonia ice. The ices are present because the cold temperature in the upper atmosphere can reach 250 degrees Fahrenheit below freezing.

The cloud bands vary in color from reds, oranges, yellows, tans, and whites. The light-colored bands are called **zones**, and the dark-colored ones are known as **belts**. No one is sure why Jupiter's clouds appear in such a

The bands and circles on Jupiter are cloud formations that fill the planet's upper atmosphere.

wide variety of colors, but some scientists think that it may have to do with small traces of elements present in the atmosphere. **Sulfur** compounds, for example, display different colors in chemical reactions. If such compounds were reacting with other elements in the atmosphere the multicolored effect may be the result. The colors, however, do show the level of the visible clouds. The red bands are the highest layers, followed by the brown, tan, yellow, and white zones. The blue cloud belts are at the lowest visible level.

The bands are formed by the strong winds that blow through Jupiter's atmosphere. The winds in nearby belts and zones blow in opposite directions at speeds up to 400 miles per hour. Scientists used to believe that the sun's heat moved Jupiter's clouds as it does on Earth. Recent findings, however, point to internal heat from the planet as the cause of the rapid wind speeds.

Lightning and Life

In addition to high winds, Jupiter's cloud layers are sometimes streaked with supercharged lightning bolts. According to astronomers, lightning is less frequent on Jupiter than it is on Earth, but the bolts are more powerful. Lightning storms are more common toward the

A close-up view of lightning storms on Jupiter, shown as small, white spots.

poles of the planet and have been spotted by space probes that have passed near the planet.

Lightning is an important energy source in driving the chemical reactions to create life on a planet. Some scientists believe lightning provided the spark that helped bring about life on Earth. Because of the presence of lightning and regions of the cloud layers where the temperatures would be livable, scientists do not rule out the possibility that some sort of life form could exist on Jupiter. The high levels of radiation in Jupiter's atmosphere and the furious cloud movements, however, argue against any form of life on Jupiter that would resemble life on Earth.

Storms on Jupiter

Lightning and high winds often signal storms on Earth, and Jupiter shows signs of storm activity as well. Amid the fast moving cloud

Small circular and oval shapes may be the cloudy tops of whirlpool storms.

The Cloud Planet   17

bands, smaller circular and oval shapes are present. No one is sure what these regions are or what has caused them, but scientists assume they are the cloudy tops of whirlpool storms below the visible surface.

Some of these storms arise and fade fairly quickly, but others have remained on Jupiter's surface for hundreds of years. Three white oval shapes in the southern hemisphere, for example, have been observed since the 1930s when they broke through the brown cloud bands. Even before that, similar—or perhaps earlier versions of the same—storm tops were seen in this region. There seems to be no way to predict exactly when or where these storms will arise and subside.

The Great Red Spot

The three oval storm shapes share the southern hemisphere with Jupiter's most well-known, the Great Red Spot. The Great Red Spot is another cloudy whirlpool of tremendous size. It is more than twice the size of Earth and has kept its shape for centuries. Its color, however, has varied over the years. Sometimes it is bright red; other times it fades to pink.

Like the smaller cloud whirlpools, the Great Red Spot is the top of an atmospheric

Jupiter's Great Red Spot is more than twice the size of Earth.

storm, though this gigantic one would be of hurricane strength. The winds in the storm blow counterclockwise, suggesting to scientists that forces under the cloud layers are pushing up toward the surface. This led some astronomers to believe the Great Red Spot was caused by heat escaping from the planet's center. Still, no one is sure what is causing such a storm, nor why it has lasted for so long while other smaller storms have come and gone.

The outer edges of the Great Red Spot move in a counterclockwise motion. They complete

Cloud bands at Jupiter's north and south poles move slower than the bands closer to the equator.

one revolution in about four to six days. The inner region of the whirlpool is more chaotic. The clouds in the central area move in many directions and usually very quickly. The entire storm also moves, though its motion is difficult to detect without careful study. It is bounded by strong air currents to the north and south, so it cannot move much in those directions. Instead, the Great Red Spot drifted eastward for a time in the early twentieth century, but now drifts slowly westward across the planet.

A Continually Changing Planet

As astronomers charted the movement of Jupiter's storms and cloud bands they noticed that the planet does not rotate uniformly. Because Jupiter is primarily a fluid mass of clouds, different parts of the planet spin at different speeds. The cloud bands closest to the planet's equator move the fastest while those at the north and south poles of the planet travel at slower speeds.

In fact, there is nothing uniform about Jupiter's appearance. What astronomers have observed for centuries is a planet that is always changing. The cloud bands widen and change color with no set pattern. The whirlpool storms arise and then vanish. Nothing seems constant on Jupiter except for the existence of the Giant Red Spot—and even that changes in shape and size. Of course, what interests scientists about Jupiter is that the constant change offers possibilities for new, unexpected discoveries. The huge planet has never failed to surprise and delight the many professional and casual observers who are fascinated by its mysteries.

3
The Rings of Jupiter

In the 1970s the National Aeronautics and Space Administration (NASA) launched several space probes to explore Jupiter and other more distant planets. These missions were called flybys because the probes only passed near the planets, gathering information as they sailed through space. The first of these missions, dubbed *Pioneer 10* and *11*, sent back new information on the temperatures and pressures within Jupiter's atmosphere. The second set of probes, however, revealed something completely unexpected. Unlike the Pioneer spacecraft, the two new probes, called *Voyager 1* and *2*, carried video cameras that sent NASA scientists high resolution images of the planet and its moons.

Arriving in the Jovian system in 1979, the Voyager cameras caught sight of faint glowing disks around the planet's equator. Astronomers were amazed to discover Jupiter had rings.

Jupiter's rings are much like the rings that circle the planet Saturn. Saturn's rings, however, are much wider and more visible. They have more material in them to reflect sunlight and shine brighter. Jupiter's rings are thin, being made up of mostly tiny particles of matter. They cannot be seen with earthbound telescopes because the edge of

Voyager 2 takes pictures of Neptune (top) and its moon Triton.

The Rings of Jupiter   **23**

the ring system points toward Earth. Saturn's rings, on the other hand, are visible from Earth because they are angled so that the flat sides of the bright disks can be seen. Thus, Jupiter's rings remained undiscovered for centuries because they could only be detected from space. Now, Jupiter is the fourth planet—along with Saturn, Uranus, and Neptune—known to have a ring system.

How Did the Rings Form?

Even though the Voyager missions had revealed Jupiter's rings, these space probes did not pass close enough to the Jovian system to indicate exactly how the rings may have formed. In 1989, however, NASA—working in partnership with the German government—launched another spacecraft to study Jupiter more closely. This mission was named after Galileo, the seventeenth-century scientist who first studied Jupiter with his telescope. The *Galileo* spacecraft was sent into orbit around Jupiter where it has remained ever since. In circling the planet the orbiter got close enough to some of Jupiter's innermost moons to note that small **meteors** crash into their surfaces and kick up dust which floats out into space. These dust particles are caught by the planet's gravity and its magnetic field and pulled into

The *Galileo* spacecraft was built to study Jupiter more closely.

the ring system. The particles do not stay in the rings for long, however. The planet's strong gravity pulls the dust particles toward the planet. Since the rings remain visible, astronomers conclude that meteor activity is fairly common around the Jovian moons, creating a steady supply of dust.

How Many Rings?

The *Galileo* orbiter also uncovered another interesting fact about the ring system that the Voyager probes could not. The images Voyager sent back showed three distinct rings around Jupiter: the inner halo, the main ring, and an outer ring. The inner halo is the ring that is closest to the planet. It extends in an oblong shape from about 57,000 miles to 76,000 miles from Jupiter's center. It is formed from particles falling toward Jupiter from the second, or main ring. The main ring is the brightest of the set, and it stretches from the halo ring's boundary to about 80,000 miles from Jupiter's center.

The outer ring, called the Gossamer ring because of its light, thin appearance was at first thought to be one ring from Voyager's images, but the *Galileo* orbiter's closer inspection proved that it was one ring embedded within another. The innermost of the twinned

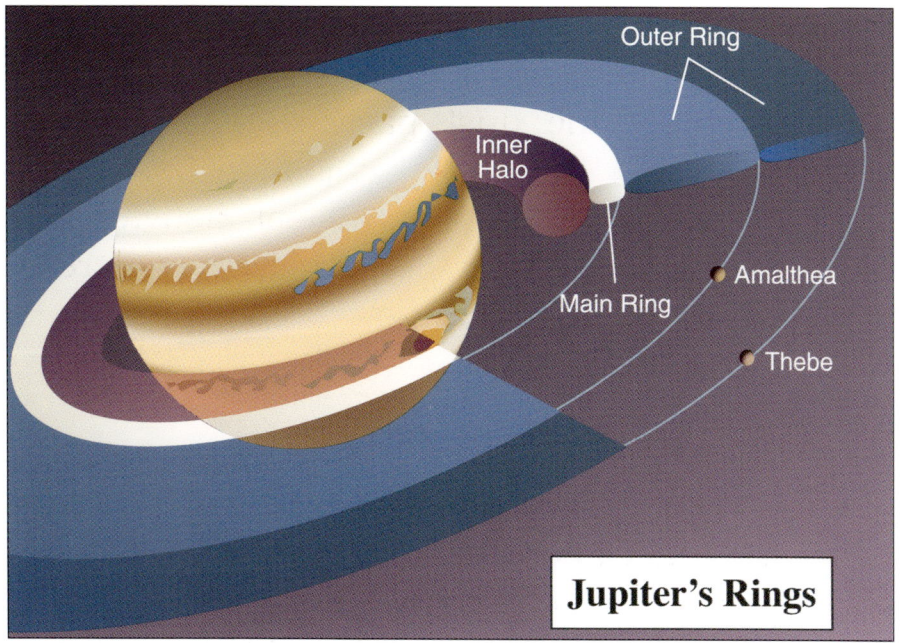

pair is called the Amalthea Gossamer ring because it extends from the main ring to the orbit of Amalthea—one of Jupiter's moons—roughly 112,000 miles from Jupiter's center. The outermost ring is called the Thebe Gossamer ring because it reaches the orbit of Thebe—another of Jupiter's moons—at about 136,000 miles out.

Continuing Investigations

Scientists still don't know everything there is to know about Jupiter's rings since *Galileo* is the only spacecraft to get close to the ring system. The *Galileo* orbiter is still circling

The Galileo orbiter sent back many images, helping scientists to better understand Jupiter.

Jupiter, lasting longer than many NASA officials predicted. It has sent many images of the planet and its rings back to Earth. It was even in the right position to record the crashing of the Shoemaker-Levy 9 **comet** into Jupiter's far side in July 1994. The comet caused giant plumes of fire to erupt in the atmosphere as it plunged into the cloud layer. Perhaps more importantly, the orbiter has passed near several of Jupiter's moons, sending back revealing pictures and data of these relatively unexplored bodies.

In October 2000, the *Galileo* orbiter received some company in the Jupiter system. Another space probe called *Cassini* passed by Jupiter on its way to study Saturn. It was a historic event because it was the first time two such probes were sending back images and data about the Jovian system at the same time. As the two spacecraft are investigating the planet, its rings, and its moons, scientists are hopeful that other unexpected aspects of the planet will be revealed.

4
The Moons of Jupiter

Although the space missions to Jupiter have found it difficult to probe the depths of the cloudy planet itself, they were able to shed light on many other aspects of the Jovian system. Most importantly, the Voyager and Galileo spacecraft sent back wonderful images of some of Jupiter's moons. These images have not only revealed much about the moons, but also about the complex relationship between these moons and their parent planet.

Just as Earth's moon orbits Earth, Jupiter has seventeen identifiable moons, or satellites—the general term for any object that orbits a planet. At least eleven more satellites were recently discovered but have yet to be confirmed by scientists. Thirteen of Jupiter's known satellites are small and appear to have been captured by the planet's gravity over time. The four largest satellites are known as the Galilean moons because it was Galileo who first noticed them with his telescope in 1610. Scientists believe these moons were created at the same time the planet itself was formed.

Ganymede

The largest of the Galilean moons is called Ganymede. At more than twice the size of Earth's moon, Ganymede is also the largest moon in the entire solar system, even bigger

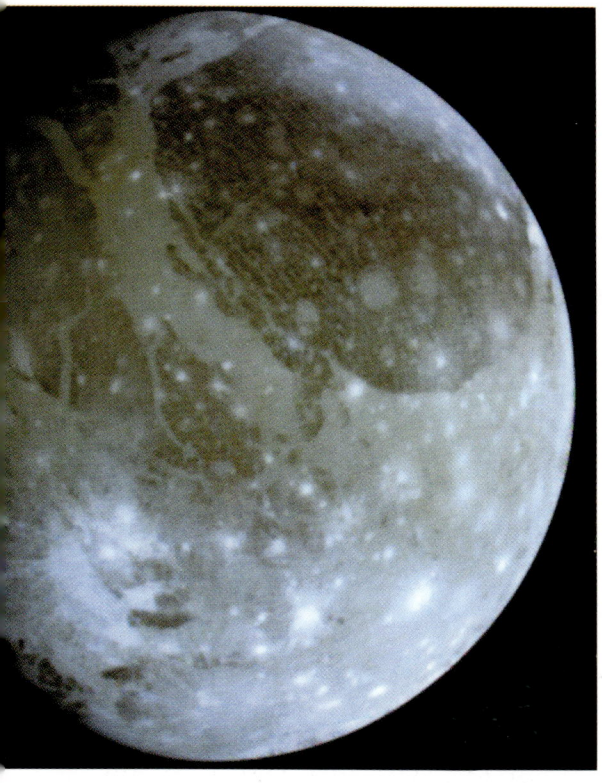

Ganymede is Jupiter's largest moon and the largest in the solar system.

than the planet Mercury. From a distance, Ganymede appears as a smooth ball covered in dark and light patches of color. The dark areas are rock and the light regions are ice. Up close, the moon's surface reveals many ridges and valleys. The ridges are usually long stretches of ice that probably formed millions of years ago when water burst up from the core of the newly created moon and froze.

High resolution images from the Voyager spacecraft also show a large, dark, heavily cratered region of the moon. The craters suggest to scientists that this part of the surface is very old. All of the Galilean moons have been hit by meteors and other debris, but many of the craters have been covered over by either ice or other material from inside the moons. The large number of craters on Ganymede's surface indicate that the moon has long been inactive and that the pockmarked landscape is the result of millions of years of bombardment.

Callisto

Callisto, the outermost of Jupiter's moons, is almost entirely covered with craters. Some of these are close to a thousand miles wide. This may seem surprising considering the moon's diameter is only about three thousand miles. Callisto's two largest craters were formed by

Callisto is almost entirely covered in craters, some being a thousand miles wide.

waves, which helps explain their size. Meteors crashed into the icy surface of the moon with enough force to melt the ice in a large region. Just as a stone dropped into a pool of water will send out several ringlets of waves, the meteor sent out ripples in the water that came from the melted ice. Since there is little rock on Callisto's surface, the waves fanned out for hundreds of miles. But the water on Callisto quickly froze again, leaving the ripples visible as ledges of ice. Other craters dot these large "bull's eye" features, giving Callisto a unique look when compared to the other Jovian moons.

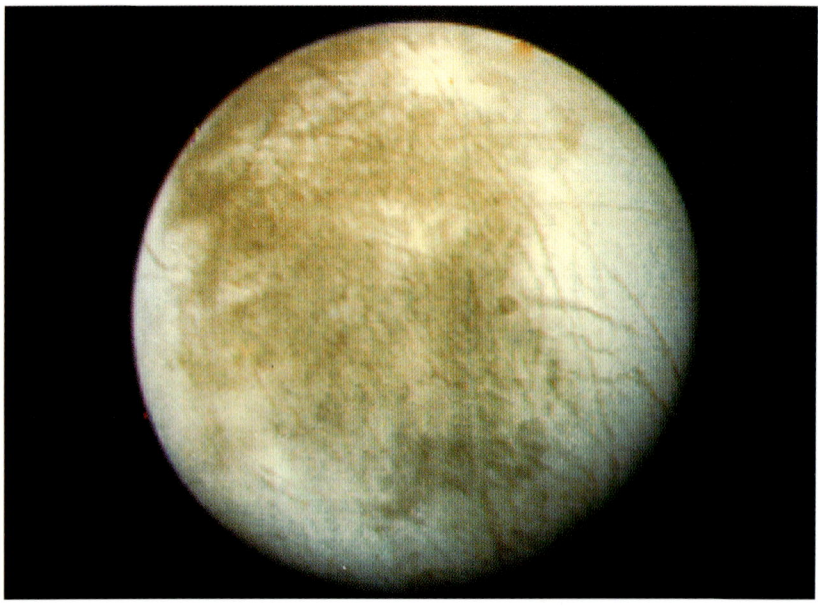

Europa is almost entirely covered with ice, giving it a smooth appearance.

Europa

Europa, the third Galilean moon, is about the size of Earth's moon. Like Ganymede and Callisto, Europa is an icy place. The surface is almost entirely covered with plates of ice, giving the moon a smooth appearance. The icy surface, however, is crisscrossed with dark streaks that may be rocky material pushed up from below.

Although astronomers once believed Europa's ice layers ran fifty or sixty miles deep, recent images from the *Galileo* spacecraft show evidence of some of the ice on the surface breaking up, suggesting that a large ocean of water may lie less than a half mile under parts of the moon's surface. The presence of water interests many scientists because it may signal the possibility of some life existing within the depths of that ocean.

Europa also shows fewer craters than the other moons which may indicate that its surface is or has been more "active" in recent centuries. That is, forces within the moon may have moved the icy sheets enough to cover over and erase many craters.

Io

The final Galilean moon, Io, is a little larger than Europa. It is an "active" moon. Unlike any of the other Jovian moons—or any other moon in the solar system—Io has several active volcanoes. *Voyager 1* was the first space probe to notice this activity. While it passed near enough to the moon to send back detailed images, it saw eruptions that sent hot material almost two hundred miles upward. Such forceful eruptions told scientists that there was great energy still trapped in Io's core.

The material spewed from Io's volcanoes contains sulfur which tinges the moon an overall red color. Sulfur compounds, however,

Io is the only volcanically active moon in the entire solar system.

display many different colors, so regions of the surface near the volcano often change color as the sulfur spreads and cools. Because Io does not seem to have any craters, scientists estimate that the volcanoes have deposited enough sulfur ash across the moon to cover over any traces of impact by meteors. Sulfur dioxide gas from the volcanoes also contributes to another rare feature on Io. The moon has a thin atmosphere. Of the sixty-one moons in the solar system, only four are known to have atmospheres. Ganymede is another one of the four.

The Smaller Satellites

Besides the Galilean moons, the only other Jovian satellite that was close enough to the path of the Voyager spacecraft to allow surface features to be seen was Amalthea. It has an oblong shape and its surface is pockmarked with craters. Io's volcanoes have sent enough sulfur into space that much of it has covered Amalthea with a reddish coating.

Amalthea is one of the four small satellites closest to Jupiter. Metis, Andrastea, and Thebe are the other three. All four have weak gravity, so tiny meteors are easily able to kick dust from these satellites into space. That dust supplies Jupiter's ring system with its reflective matter, ensuring the rings will be visible for a very long time.

The nine remaining known satellites are the farthest from the planet. They are most likely asteroids or other space debris that were captured by Jupiter's gravity when the planet was forming. The nine satellites form two groups: Ananke, Carme, Pasiphae, Sinope, and S/1999 J1 travel in one direction around the planet, while Leda, Himalia, Lysithea, and Elara orbit in the opposite

Jupiter may hold the key to understanding much about the solar system.

direction. The tiny S/1999 J1 was discovered in 1999 just outside the orbit of Sinope. Its presence has prompted scientists to look for even more Jovian satellites. In 2000, eleven more satellites were spotted, but their existence has yet to be confirmed by astonomers worldwide.

As astronomers seek to confirm the existence of new satellites, they also continue to gather information on the Galilean moons and the planet Jupiter itself. With space exploration still in its infancy it is difficult to know when—if ever—Jupiter's mysteries will be resolved. The big cloudy planet, so different from Earth, may hold the key to understanding much about the entire solar system. But until humans can unlock its secrets, Jupiter, the "lord of the sky," will maintain the interest of astronomers and casual observers alike.

Glossary

ammonia: A gas formed by combining the elements nitrogen and hydrogen. Ammonia exists as a gas and as a solid ice in Jupiter's atmosphere.

belts: The name given to the narrow dark-colored bands of clouds on Jupiter.

comet: A heavenly body with a starlike center and a long luminous tail. Comets orbit the sun and sometimes crash into planets and moons. The Shoemaker-Levy 9 comet hit Jupiter in July 1994.

gravity: The force of attraction. Gravity on Earth draws all matter to the ground. Jupiter has no visible "ground" but gravity still pulls everything toward the planet's center.

helium: A common element. Helium is one of the principle gases in Jupiter's atmosphere. Because it appears in sunlike proportions, scientists think Jupiter may have nearly become a star as it formed.

hydrogen: Hydrogen is a very plentiful element in the solar system. It is the most abundant gas in Jupiter's atmosphere. Many astronomers also believe hydrogen exists in a liquid form below Jupiter's cloud layers.

magnetosphere: The magnetic field that surrounds many heavenly bodies. Jupiter's magnetosphere pulls charged particles from space into its ring system. The high number of charged particles also accounts for a high level of radiation around the planet.

meteor: A heavenly body with a solid, rocky form. Large and small meteors travel through space and sometimes collide with planets and moons. Jupiter's moons, for example, have been hit many times by meteors, leaving visible impact craters on their surfaces.

methane: Another one of the gases present in Jupiter's atmosphere. It is a compound of carbon and hydrogen.

orbiting: Heavenly bodies revolving around another. The revolving body is held in orbit partly by the gravity of the central body. This helps explain why an orbiting body continues to revolve around the central body instead of flying off into space.

solar nebula: A swirling cloud of gases from which some scientists believe the sun and planets formed.

solar system: The sun and the nine planets, moons, and other smaller planetoid objects (such as comets) that revolve around it. Jupiter is the fifth planet from the sun.

sulfur: Another common element. It can be found on both Jupiter and its moons. Sulfur interacts readily with other chemicals to form compounds. These compounds exhibit varying hues, which explain some of the colorful surfaces seen in the Jovian system.

zones: The name given to the wide light-colored bands of clouds on Jupiter.

For Further Exploration

Books

Reta Beebe, *Jupiter: The Giant Planet*. Washington, DC: Smithsonian Institution Press, 1994. One of the more current books about the planet. Although a complex study of the planet, a lot of useful, basic information can be found here.

Gary Hunt and Patrick Moore, *Jupiter*. New York: Rand McNally, 1981. An informative and easy to follow book on the planet and its moons. The presentation of information is thorough, ranging from basic concepts to more technical discussions. Color and black-and-white photos and illustrations are spread throughout.

Patricia Lauber, *Journey to the Planets*. New York: Crown, 1993. An introductory guide to the planets and space exploration. This book covers what we know about heavenly bodies and how we know it—from the landing on the moon to probes to other planets.

Websites

NASA Galileo Project

www.jpl.nasa.gov/galileo

NASA's official website concerning the Galileo space mission. Recent images from the probe as well as press releases provide an excellent resource. This is the best source for up-to-date information on Jupiter and its visible moons.

NASA Kids

http://kids.msfc.nasa.gov

This is the space organization's website designed especially for kids. There is very basic information here on the planets and many fun projects to try. The site is updated regularly with reports on the best times to view astronomical events as well as upcoming NASA projects.

Index

Amalthea (moon), 27, 37
Amalthea Gossamer ring, 27
ammonia, 8
ammonia ice, 13–14
ammonium hydrosulfide, 13–14
Ananke (satellite), 37
Andrastea (satellite), 37
atmosphere
 gases in, 8
 of Io, 36–37
 upper, 13–16
belts (of color), 14
Callisto (moon), 32–33

Carme (satellite), 37
Cassini, 29
cloud bands
 described, 13–15
 rotation of, 21
 storms in, 17–18
color bands, 13, 14–15
comets, 28–29
dust, 7–8
Earth
 clouds on, 13, 15
 lightning on, 15, 16
 location of, 5
 radiation belts of, 11
Elara (satellite), 37–38
equator, 10, 21
Europa (moon), 34–35

Index 45

flyby probes, 22, 29, 35
Galileo, 24, 26–29, 34–35
Galileo Galilei, 6–7
Ganymede (moon), 31–32, 37
gases, 7–8
gas giants, 6
Gossamer rings, 26–27
gravity, 6
Great Red Spot, 18–21
heat
   amount of, 8
   Great Red Spot and, 19
   movement of clouds and, 15
helium, 8
Himalia (satellite), 37–38
hydrogen, 8, 11–12
ice
   Callisto and, 33
   cloud layers and, 13–14
   Europa and, 34
   Ganymede and, 32
Io (moon), 35–37
Leda (satellite), 37–38
life, 16–17, 35

lightning, 15–16
location, 5
Lysithea (moon), 37–38
magnetospheres (magnetic fields), 10–11
Mars, 5
mass, 6
Mercury, 32
meteors, 24, 26, 32–33
methane, 8
Metis (satellite), 37
moons, 7
   Amalthea, 27, 37
   of Earth, 5
   Galilean
      Callisto, 32–33
      Europa, 34–35
      Ganymede, 31–32, 37
      Io, 35–37
   Galileo and, 29
   meteors and, 32–33
   number of, 30–31
   smaller, 37
National Aeronautics and Space Administration (NASA), 22, 24, 26–29
Neptune, 6, 24
nuclear reactions, 7–8

Pasiphae (satellite), 37
Pioneer 10 and *11*, 22

radiation belts, 11, 17
rings
  dust and, 37
  formation of, 24, 26
  Gossamer, 26–27
  number of, 26–27
  visibility of, 23
Romans, 4
rotation, 9–10, 21

satellites, 38–39
  Amalthea, 27, 37
  of Earth, 5
  Galilean, 31–37
  Galileo and, 29
  meteors and, 32–33
  number of, 30–31
  smaller, 37
Saturn, 6
  magnetosphere of Jupiter and, 11
  rings of, 23
Shoemaker-Levy 9 comet, 28–29
Sinope (satellite), 37, 38
size, 4, 5
S/1999 J1 (satellite), 37, 38
solar nebula, 8–9
solar system, 4–5

space probes, 22, 24, 29, 35
stars, 7–8
storms
  described, 17–18
  Great Red Spot, 18–21
sun, 5, 8–9

telescopes, 6–7, 23
temperatures
  life and, 16–17
  in upper atmosphere, 14
Thebe (moon), 27, 37
Thebe Gossamer ring, 27

upper atmosphere
  cloud bands in, 13–15
  colors in, 14–15
  lightning in, 15–16
  temperatures in, 14
  winds in, 15
Uranus, 6, 24

Venus, 5
volcanoes, 35–36, 37
Voyager 1 and *2*, 22–23, 35, 37

water, 13–14
water vapor, 8
winds, 15

zones, 14

Picture Credits

Cover & Title Page Photo: PictureQuest
© AFP/CORBIS, 16, 31
© Bettmann/CORBIS, 8, 20
© CORBIS, 17, 23, 25, 38
© Araldo de Luca/CORBIS, 5
Chris Jouan, 10, 12, 27
Chris Jouan and Martha Schierholz, 7
NASA, 9
Courtesy of NASA/JPL/Caltech, 14
© NASA/Roger Ressmeyer/CORBIS, 33, 34
© PhotoDisc, Inc., 19, 36
© Reuters NewMedia Inc./CORBIS, 28